Download The Audio Version Of This Book Free!

If you love listening to audiobooks on-the-go or enjoy the narration as you read along, I have great news for you. You can download the audiobook version of

Capsule Craze:
The Comprehensive Guide to Building a Capsule Wardrobe for FREE

just by signing up for a FREE 30-day Audible trial.

Use the links below:

For the Audible UK: https://tinyurl.com/y2rzgzc8
For the Audible US: https://tinyurl.com/y3fe7whd

REVIEWS

Reviews and feedback help improve this book and the author. If you enjoy this book, we would greatly appreciate it if you could take a few moments to share your opinion and post a review.

Capsule Craze

The Comprehensive Guide
to Building a Capsule Wardrobe

Rebecca Ellington

ISBBN 978-1-7350253-5-3

Table of Contents

CHAPTER 1

Introduction

What jAre Capsule Wardrobes?

Here you are again. It's Friday night, and you're standing in front of your closet, blankly staring at the pieces on your hangers, a sense of panic building as you try desperately to make a decision on what to wear. The sweater you wore earlier is much too heavy. You've got a dress that's classic and comfortable, but you don't have any heels to match. On your bed is a massive pile of rejected outfits that you have tossed aside in frustration. And accessories? Forget about it. You know you should have gone to the mall and found a new trendy outfit, but that would require time and money you just don't have.

Your date is waiting at the front door, growing more and more impatient by the minute. Why, oh why, is it so hard to find something to wear?

The solution is easier than you think. If this scenario sounds all-too familiar, consider a high functioning, easily adaptable, **Capsule Wardrobe.**

Oh, you haven't heard? Capsule wardrobes are IN! As a matter of fact, capsule wardrobes have been a fashion mainstay for over thirty years. With a focus on interchangeable, long-lasting garments, capsule wardrobes will likely be IN as long as fashion is a must!

If you're unfamiliar with the term, then you're in the right place.

Most people consider clothing pieces "old" after wearing them three times. In some cases, as with wedding or party dresses, pieces are only worn once. Additionally, the "Fast Fashion" trend has swept the world, with technology allowing quick manufacture of inexpensive clothing. Trends are evolving more and more rapidly, meaning what walked down the runway last month is no longer in style. Clothing has become less sustainable and more disposable at the cost of the planet and your wallet! This cultural trend has manifested itself in compulsive buying; therefore, clothing retailers and manufacturers have to keep costs affordable for the consumer by using lower quality materials and less expensive labor. This one-time-use, disposable "fast fashion" concept has resulted in more and more clutter and waste.

At the end of the day, the "fast fashion" culture has made women believe that they need to have a closet chock full of options ready for any pending situation, when in reality, having more choices at one's disposal leads to fashion anxiety. Symptoms of fashion anxiety include indecisiveness, stress, and the feeling that you have absolutely nothing to wear!

Enter the capsule wardrobe! As a cure for fashion anxiety, capsule wardrobes are trend-driven designers' greatest nightmare. This secret weapon incorporates reusable, interchangeable, and high-quality staple pieces that can be combined in different manners for any and every occasion. Women can refine their cluttered wardrobes to include only high-quality essential pieces that can be worn and re-worn in a variety of ways.

The idea is to fill one's closet with timeless pieces that can be mixed and matched to create a plethora of outfits. If you've created your capsule correctly, you should need approximately thirty pieces in your wardrobe for regular wear.

Think about how many outfits that are currently hanging in your closet right now. Do you think you can reduce that to thirty pieces?

Take a deep breath! With this comprehensive step-by-step guide, you'll learn how to do exactly that.

CHAPTER 2

History

Where Did Capsule Wardrobes Come From?

In the 1970s, fashion choices were bold, empowering, and verging on strange. Fashion icons like Cher, Joni Mitchell, Bruce Lee, and even Jack Nicholson all received creative recognition for a variety of styles, including flared-out pants, tie-dyed shirts, wide-lapeled suits, platform shoes, and fringed suede jackets. Disco was king, and wild fashion was queen. It was a time in which clothing begged for attention, and fashion designers of the day were free spirited with an "anything goes" attitude.

Meanwhile, in a small London boutique called *Wardrobe*, owner Susie Faux was igniting a new trend that stood out in stark contrast to the outlandish styles that were flooding the streets. While developing the next big wave in the fashion industry, she couldn't have predicted that she was on the verge of discovering a long-term fashion solution for the ages.

In all the hustle and bustle, Susie Faux saw the necessity for a different approach to fashion—one that was simpler, more subdued, and interchangeable. Like many cities, London experiences drastic seasonal changes, so having a flexible, multi-purpose wardrobe is essential. It is with this in mind that she developed and coined the term "capsule wardrobe."

The idea behind the boutique known as *Wardrobe* was to create an all-in-one shop of essentials, where professional women could shop with confidence. Here, they could find every item they might require in a single, stand-alone location. Susie Faux would become the "godmother" of classic fashion innovation, as her clients quickly became trusted friends and loyal customers. Her formula was simple and perfect for the tumultuous time: she sold staple pieces and inspired women around her to simplify their wardrobes and their lives.

As the daughter and granddaughter of two master tailors, Susie Faux's eye for high-quality fashion was ingrained from an early age. In adulthood, she became a respected member of the London fashion community. She quickly became known for introducing British women to up-and-coming designers that would soon become international sensations, including Gianfranco Ferre and Jil Sander.

The concept of the capsule wardrobe was later popularized by American designer Donna Karan, who is touted as being the woman who launched the concept of capsule wardrobes beyond London and into the fashion mainstream around the world. In the mid-1980s, she released an influential collection of seven interchangeable work-wear pieces, and the capsule craze took off.

Today, the term "Capsule Wardrobe" is continually referenced in newspapers, magazines, and fashion editorials, proving it has achieved a level of fame that many designers and fashion innovators can only hope for. As for Susie Faux, she is regularly complimented for being the woman who made it all possible.

Susie Faux

CHAPTER 3
The Benefits of Capsule Wardrobes

There are many benefits to reducing the footprint of your wardrobe, and some of them are not as obvious as simply clearing the clutter.

Have you ever noticed how self-made millionaires and innovators are always caught wearing the same thing? Their signature look becomes their brand. When someone mentions Mark Zuckerberg, for example, it's easy to envision the curly-haired technology guru in his casual hoodie sweater, a pair of jeans, and lightweight sneakers. How about Apple founder Steve Jobs's turtleneck sweaters? Even former President Barack Obama admitted that he limited his suit options to those of the black and blue variety.

What those successful people have realized is a concept the rest of us are slowly catching up on: reducing the amount of time your brain spends mulling over insignificant actions like dressing allows your mind to be open and refreshed for the more complex decisions in our day-to-day lives.

Simply put, less is more when it comes to fashion decisions. But how?

It Reduces Decision Fatigue

"Decision fatigue" is the inevitable deterioration of quality decisions made by individuals who have been participating in long sessions of decision making. It's something everyone experiences, even those who work in high-stress roles where tough decisions are made all the time. For those who make a lot of tough choices at work, at home, while parenting, or even in play, it is possible to experience decision fatigue every single day. If you've ever experienced that 2 p.m. afternoon crash at the office, you may be experiencing the exhaustion that comes with the stress of making a lot of decisions throughout the day.

By reducing the overall amount of daily decision making, mental clarity is gradually enhanced. Take away some of the stress, and the bigger picture comes into view. The result of reducing decision fatigue can be more productive days where you can be confident in making decisions that are not being clouded by stress and exhaustion.

It Requires Less Time

Consider these statistics: the average woman spends 16 minutes every weekday morning trying to decide what to wear, and about 14 minutes every weekend morning. If that doesn't sound like a considerable amount of time to you, then apply those figures to these scenarios:

- midday outfit changes when running from the office to dinner, or to the gym

- date nights and parties
- unexpected accidents, like spilling a beverage on a blouse, thereby prompting an immediate outfit swap.

All in all, women spend 287 days out of their entire life thinking about what to drape over their bodies. Just think about what they could do with all that extra time if they could reduce their overall decision-making time by creating a more limited, yet more flexible, wardrobe.

This stress is not exclusive to women, either. Think about children getting ready for school, having to keep in mind the activities of the day when planning what to wear. Men also have big decisions to make when it comes to building an outfit, matching separate pieces to create an ideal look for whatever the day may bring. Just consider the amount of time your household may spend just digging around in drawers and closets, trying to pull together something suitable to wear before leaving the house. Wouldn't it be great to use that time for something more pleasant?

It Reduces Stress

Fumbling about in your bedroom as you struggle to determine if your outfit is too formal, too casual, too uncomfortable, too short, or too long for whatever event you're about to head out for is obviously very stressful. Perhaps this decision-making process is making you run behind schedule. Perhaps you have to skip steps like fixing your hair, showering, or running a last minute errand because of the extra time spent on picking out an outfit. Being frequently stressed reduces overall self-esteem as your decision-making ability is being challenged. Given that stress has been shown to increase the risk of obesity, depression, and heart disease, it's clear that limiting our sources of stress is highly beneficial to our overall health.

It Wastes Less Energy

Not only do large, complicated wardrobes require more detailed decision making, they also require more upkeep and maintenance and more attention to keep all those pieces organized. Think about all the time you spend just shuffling your summer clothing to the back of the rack to make your fall and winter styles more accessible and apparent as the seasons change. Many people block off an entire day to "switch seasons" with their clothing. Wouldn't it be great to have that entire day back?

Also, think about how laundry days will change, when you no longer have to wash, dry, press, fold, and sort load after load of separate pieces that you may only wear once a month. Having fewer pieces overall means less time spent sorting through your laundry bins as you prepare them for a wash and less time afterward as you fold and press them for storage. It also reduces the number of musty, forgotten pieces accumulating in the bottom of the hamper or hanging in the laundry room, waiting patiently for a trip to the cleaners that will never come.

It Inspires Confidence

When you force yourself to create a more focused wardrobe, you're forcing yourself to take a good, long look at your style and the kind of pieces that not only make you look good but feel good too.

Think of your closet as a luxurious five-star restaurant: the menu might only be a single page, but rest assured, every single ingredient and spice was chosen with a purpose —to create dishes that are amazing and satisfying in every way. Compare that to when you visit a large family-friendly chain restaurant where, while the possibilities are seemingly endless, they are less creative, lower quality, and generally uninspired. When you're a victim of "Fast Fashion," your wardrobe starts looking less inspired and intentional and more generic and inexpensive. As a result, you can find yourself feeling like nothing fits right or looks good enough to bother wearing.

Building a sustainable capsule wardrobe is an exercise in choosing pieces deliberately with the intention of making yourself look and feel great in a variety of settings. No longer will you grab quick pieces and "make them work." Instead, you'll buy pieces that fit your body, your style, and your lifestyle needs. Think of how many times you've picked at a hem, dug at a collar, or wriggled around in a waistband that wasn't quite right. Think of the times that you've left the house feeling frumpy or embarrassed at how your body feels in your clothing. Wouldn't it be great to never feel like that again?

When you're sure of your style, you're sure of yourself, and that confidence will manifest itself throughout additional parts of your life, including in the relationships you have with your family, your friends, and in making decisions at your job.

It Saves Money

We touched earlier on how rapidly changing fashionable trends and cheap manufacturing methods have created a culture of fast fashion. While this can be helpful, as it makes buying new pieces affordable for the average consumer, it is not without its shortcomings. When those pieces can be worn only once or twice due to trend, durability, or necessity, how much are you really saving? The answer is not much at all. By creating a small collection of timeless essentials, you'll save upwards of 75 percent of your current shopping budget, leaving you with significantly more financial freedom.

As you can see, building a capsule wardrobe has the potential to improve your life in many ways. Not only can this system reduce your overall stress, you'll also have the opportunity to reclaim your confidence levels and make your day-to-day choices more manageable. You'll save time and considerable amounts of money by making it easier to fine-tune your style, and before long, strangers in the street will be wondering how you became such a fashion-forward success.

CHAPTER 4

Questions to Ask Yourself Before Building Your Capsule Wardrobe

At this point, you're probably pretty excited about your new capsule wardrobe, but you may still be wondering if it's something that would benefit you on a daily basis. To help you determine if you should take the capsule plunge, there are a few questions you should ask yourself. After all, it's that impulsive nature that led you to a closet full of impossible choices, so be prepared before you make the switch!

1. **Do you frequently have difficulty finding something to wear?**

    You know what we're talking about: standing in front of your stacked-to-the-brim closet as that feeling of dread creeps upon you, realizing you have absolutely nothing to wear! Having too many options can completely overwhelm our senses, leading to stress, anxiety, and a strong urge to run to the mall for binge shopping. If you find yourself in this predicament, it's time to declutter that closet and make room for a capsule.

2. **Would having more time in the morning benefit you?**

    For many people, every minute counts. Would having an extra ten or twenty minutes in the morning make starting your day a little easier? Taking the difficult decision making out of the equation can add up big time, and a capsule wardrobe could be the quick fix you need.

3. **Do you want to save some money?**

    If your credit card bill is spinning out of control while you try to keep up with the latest trends, consider reprioritizing your clothing budget to include high-quality, practical pieces that can be interchanged throughout the year. Streamlining your closet could help you get a handle on your finances, allowing you to use funds for something more exciting, like that dream vacation you keep telling yourself you'll take.

4. **Are you tired of putting on old clothes that don't compliment your personality, body type, and style?**

    When you have a closet full of clothes, it's easy to forget what you have lurking at the back of the rack. Before long, you'll find that some of the "lost" pieces no longer fit, are no longer in style, or simply don't represent you anymore. Sometimes, these neglected items will be worn through with holes or pills from overwashing, making them completely unwearable.

That's one of the downfalls of fast fashion and why capsule wardrobes focus on well-made clothing that is built to last. By choosing essentials that'll complement your shape, your style, and your comfort level, you're far less likely to buy something you don't enjoy wearing. If you want to regain confidence in yourself and your style, capsule wardrobes can help.

Consider this: all of your capsule wardrobe items will be carefully chosen by you, for you, based on how they make you feel. No longer will you grab something trendy from the mall, hours before your date, and hope that it fits right. Have you ever spent what feels like an eternity standing in front of a mirror, pulling at a hemline or picking at a neckline in paranoid horror, hoping that it falls just the right way? When you take the time to choose a fit, fabric, and style that feels just right, you say "farewell" to the picking and the paranoia. Capsule wardrobes are all about choosing what you feel good wearing all the time, and as you read on, you will discover how to choose items that you can feel 100 percent confident wearing, 100 percent of the time.

5. **Do you worry about the footprint your fast fashion choices is creating?**

In order for manufacturers to churn out many new styles regularly throughout the year, compromises are often made with regard to the quality of the clothes. Additionally, there can be some ethical concerns in how factory workers are compensated and treated. Taking control of your fashion purchases can make an impact globally, and minimizing your wardrobe can help you focus on brands and designers you believe in. Creating a capsule wardrobe is a gradual process, rather than a snatch-and-grab effort at a local retail location. You are encouraged to research brands and designers to make sure their ethics and sustainability efforts align with your own. As you read on, you'll learn ways to further decrease the global footprint of the clothes you will add to your capsule and the clothes that will be removed from your cluttered collection.

If you answered yes to any of the above questions, it's time to start talking about how to structure your first capsule wardrobe.

CHAPTER 5

The Structure of Your Capsule

Building a framework to work from is the ultimate key to a successful capsule wardrobe. Your ideal capsule should consist of a small selection of carefully selected pieces that help you express your style while being functional for your lifestyle. Remember: the end goal is to be able to quickly grab your clothes and go on any given day.

To help you determine the structure of your capsule, we've outlined the four key areas of consideration for your framework.

Step 1 Outline the Concept: Finding Your Own Style

First, you're going to want to look ahead to the coming season and ask yourself what you want your style to look like. Do not panic! Take a deep breath, and get ready to use your imagination. This is a fun exercise that forces you to forget about what you already wear. Instead, think about what you want to wear. Most of us will probably say "yoga pants and a T-shirt," but really dig deep here. Grab a pen and paper, because this is going to be an exciting brainstorming session!

Think of a time when you felt really good about your outfit. Maybe this was at an event. Maybe this was just a random day at work. Think of how your body felt. Think of the compliments you received from others. Now jot down what you were wearing.

Next, think of some looks that you really like. These can be looks that you've seen on other people, in magazines, on television, etc. At this stage, include everything that strikes your fancy. Don't discount a look just because it might be expensive, you can't find a picture in the right color, or you're not sure how to work that specific piece into your overall style. We'll get to that later.

You'll also want to make note of things that make you feel good. For example, if it's turning to fall and you know you love to wear high boots, leggings, and long sweaters, include these here. Narrowing down the specifics of your likes and dislikes will save you time in the long run. If you hate the way wool feels or don't have time to dry-clean silk, this is where you want to jot that down. Knowing what you love is extremely important, so make notes of some of your favorite go-to outfits. At the same time, acknowledge what you don't love. A capsule wardrobe has no room for clothing you secretly hate, so be honest with yourself in what you write down. The best thing about a capsule wardrobe is that it is a true reflection of yourself—there's no more hiding behind fast fashion or trends that you think you "ought" to be wearing.

By now you should have a whole bunch of nebulous ideas about what you want to wear going forward, and it's probably a lot more than thirty pieces total. This is fine! Now you'll start to narrow things down. It is here that you're going to want to create a long list of the elements you hope to include in your capsule, such as colors, textures, and fabrics. This entire process is designed to help you get excited about what your finished capsule will look like, which will serve as a great motivation throughout the process

It might be beneficial to create a mood board to use to play around with different themes and ideas until you feel confident in your decision. You might even want to start a lookbook online to help you track some of the looks you like. Mood boards and lookbooks are basically a grown-up form of paperdolls, where you can compile a bunch of items that interest you, then organize them to see how they work together. Pinterest, of course, is a popular choice, but there are actually quite a few programs out there that allow you to upload fashions virtually, including phone apps that let you take your imaginary closet everywhere you go. Do a quick search for "lookbook app," and you'll find many examples. Feel free to choose one that works with your operating system—both your computer and your sense of organization! At the end of this exercise, you'll want to discover three key things to help you build your capsule: your preference in color, your favorite fabrics, and the looks you tend to choose naturally. These are the key elements in defining your own personal style.

Step 2 Create the Uniform: Bringing Your Style to Life

Once you have determined an overall concept, it's time to create one functional uniform, or outfit formula, that you can see yourself wearing. Remember how Barack Obama narrowed down his daily style to blue and black three-piece suits? This is your moment to take control of your wardrobe in the same way. This is your chance to say "So long!" to fast fashion, to outrageous credit card bills, and to all of the clothes you'll never fret about forgetting again. Think of this as the first step towards sustainability, not the last step in enjoying your clothing!

You may already have a uniform without consciously being aware of it. Look at the notes you have made, and evaluate what you have done.

1. Do you tend to reach for the same pair of jeans, despite having several pairs?

2. Do your friends or coworkers joke that you have an "addiction" to a staple item, like cardigans, vintage dresses, or sweater vests?

3. Do you put on the same sweater whenever you're feeling a little chilly?

4. Does your mood board or lookbook focus on one specific color or shape?

5. Do you currently own multiples of the same item (think of Steve Jobs and his black turtleneck)?

Answering "yes" to any of these questions helps you discover what your current uniform may lean towards. The next question to ask yourself is very important and will help shape the future of your capsule wardrobe:

IS THIS WHAT I ENJOY WEARING?

Think carefully before you answer. There is no right or wrong answer. It may very well be that you have been wearing black T-shirts out of desperation, but you secretly long for color. You may have been wearing leggings because you dreaded shopping for jeans. Now is the time to be honest with yourself, because one of the advantages of a capsule wardrobe is never having to wear something you hate, ever again.

From your mood board, lookbook, and notes, you'll start to notice some options becoming more appealing than others. Feel free to remove or cross off those items that don't fit into the uniform you are creating.

As you are diving into the process of setting up your uniform, bear in mind the practicality of the elements you choose. If your office dress code prohibits leggings, for example, don't create a uniform that's heavy on casual pieces. Alternately, if you work from home, you probably won't need a host of formal suits. This is another part of being honest with yourself and balancing needs and wants. Don't worry…we're not done yet!

Another thing to consider at this point is accessories. Don't dwell too much on these items, as we'll cover them more extensively in chapter 14, but start to think of things like:

1. What type of shoes will look good with these tops and bottoms?

2. What are your favorite jewelry pieces?

3. Do you typically wear things like scarves?

Your mind may start to wander a bit, and it's OK to make note of your current favorites, such as how you typically wear black ballet flats with everything or how you only have brown belts. These are important characteristics to incorporate into your uniform.

At the end of this stage, you should have a much more clear idea about what you can and will wear. While you won't have your wardrobe narrowed down to those thirty staple pieces just yet, you should be feeling more confident about the direction you are headed. You might be feeling the itch to purge your closet of unwanted items and inspired to rid yourself of clothing items you're confident no longer serve you.

Step 3 Build the Framework: Choosing the Individual Pieces

Now that you understand your outfit style, it's time to put together the basic essentials of a functional capsule wardrobe. You'll use your uniform, as well as the elements you identified on your mood board or lookbook, as looks you'd like to incorporate into your wardrobe.

Let's take a look at an example of how to create a framework around a uniform. Remember, this example may not pertain to your particular style, but is intended to give you an idea of the process of narrowing down the many looks you may enjoy into a functional, stream-lined wardrobe you can enjoy every day.

Uniform:
T-Shirt, Jeans, Sneakers

Identify an outfit as a uniform: in this case, a casual outfit of T-shirt, jeans, and sneakers. Move everything on your lists, moodboards, and lookbooks that fits into this uniform into one location.

Concept Elements:
Boots, Leggings, Long-Sleeved Shirts

Here, you identify some of the pieces on your list of things you love. Consider how they can be incorporated into a variety of looks.

In this example, the look was dissected into its main concepts: long-sleeved shirt, boots, and leggings. Notice how the concept pieces are pictured in the lookbook with accessories, such as a fall coat and a scarf. These are the concepts that will help you identify individual pieces of your uniform going forward.

When building the framework, you should estimate how many of each item you will require, aiming between twenty and thirty pieces in total. These can always be adjusted, so just make your best-educated guess.

Key Things to Keep in Mind When Creating a Uniform

You will want to have several versions of each item in your uniform. A good starting point is to allocate 50 percent of your overall items to the categories your uniform falls under. So, if your uniform is a T-shirt, jeans, and sneakers, and your capsule wardrobe goal is thirty pieces, you want fifteen of those pieces to be T-shirts and jeans.

Consider how often you will be reaching for the other items in your capsule and distribute accordingly. If it seems unlikely you will need to wear a dress more than once in the upcoming season, you don't need five different dresses. While the temptation to plan for possibilities may exist, remember, you're going to be choosing a dress that can handle multiple looks, so don't jump ahead of your plan. The same goes with anything else that you don't see yourself wearing as much. Pick items that you will reach for often and purchase them over items you may only wear once or twice in a year.

Items that require more frequent washing should be considered of higher importance than those that require infrequent washing. If your hoodie can survive five days without a wash, you will need fewer of them in total, as opposed to things like underwear that require frequent washing.

Regularly worn items like shoes and jackets will have a longer life and need to be replaced less often if they are rotated every couple of days; therefore, allocating at least three items to these categories will benefit you in the long run.

Structuring Your Uniform

This is the step in which you will actually chart out the pieces you think you will need and the number of pieces in total. In this example, the uniform is T-shirts and jeans, with additional concepts of long-sleeved shirts, leggings, jackets, and dresses.

As you look at this structure, the concept of the capsule wardrobe should start to hit home. This example includes a total of ten tops and six bottoms. If those are all interchangeable, that's sixty possible combinations right there. Add one of the three jackets, and you've more than doubled the combinations. The two dresses can be worn alone or with a jacket. You don't have to be a mathematician to see that there are already plenty of options ready to go with just twenty-five core items!

T-Shirts	Long sleeves	Jeans	Leggings	Sneakers	Boots	Jackets	Dresses
5	5	4	2	2	2	3	2

As you can see above, the uniform (T-shirts, jeans, and sneakers) categories have the highest quantities, while the upcoming seasonal pieces (long sleeves, leggings, boots, and jackets) come second. You can stack the quantities to favor the pieces you think you'll wear more often, depending on your lifestyle and where you live. If you really enjoy wearing dresses, for example, let yourself have a few extra pieces in that category.

Step 4 Draft Your Final Capsule

Using your basic structure guide above and the concept you visualized for yourself, get more specific about your capsule and start narrowing down colors and textures. Now is the time to get down to the nitty-gritty and really define your likes, loves, wants, and needs.

In this example, we're drilling down past the basic concepts and into the actual looks, fabrics, and colors we want in our example capsule wardrobe. Notice in the table below how different aspects are included, like style and color. You'll also want to consider how many of each item you want. Remember, this is just an example. You may prefer to have five entirely different shirts, instead of multiple identical items. Refer to your concepts and your uniform to get a feel for how frequently you'll need to reach for these particular items, and consider a realistic number of items to incorporate into your capsule.

An Example Draft:

T-Shirts	5	(2) White (2) Black scoop neck (1) Red V-Neck
Long-sleeved Shirts	5	(2) Black (2) Grey crewneck (1) Red stripes
Jeans	4	(1) Black skinny (1) Blue loose fit (1) Blue slim fit (1) Ripped
Leggings	2	(1) Black (1) Grey
Sneakers	2	(1) Black runners (1) White casual
Boots	2	(1) Black, knee-high (1) Suede ankle
Jackets	3	(1) Blue, denim (1) Windbreaker (1) Black, leather
Dresses	2	(1) Short sleeves, cotton, floor length (1) Long sleeves, lace, scooped

Drafting Tips:

1. Look at what you already own, pull pieces out, and separate the items that already fall within your capsule specifications. If your style was already well defined, you might be able to build most of your capsule with items you already own. Note: it may be tempting to want to reincorporate some forgotten clothing items back into your draft once you pull them from the closet. Put these items to the side for careful consideration. Resist the urge to completely revamp your mood board or lookbook!

2. Examine the categories you've outlined in sets to help determine a good balance of neutrals/bright colors and basics/statement pieces for optimization. If you already have your heart set on a bright yellow blazer— go for it. Just be sure to balance the rest of the colors to make most of the pieces truly interchangeable. We'll get into color palettes in more detail shortly.

3. Keep a list of your final draft available for future reference. This is where the mood board or lookbook app can be really handy. As time goes by, you'll want to keep a careful note of items that need to be repaired, replaced, or added. This can ensure your capsule remains tidy and organized, so you don't find yourself buying unnecessary pieces.

Color Palette Considerations

One of the areas in which many people struggle is in creating a color palette. After all, if your uniform is polo shirts and khaki pants, isn't it so much easier to just buy a dozen polo shirts in a dozen colors? When looking through the lens of the capsule wardrobe, however, you're looking at pieces you will really wear and can incorporate into a variety of outfits. That means the yellow polo you never actually wore is gone and in its place is an item you actually enjoy wearing.

Base Tones

It's important to spend ample time figuring out what colors you feel most comfortable wearing. Choosing which bright and bold colors you enjoy will help you determine your base tone. Think about what colors you tend to reach for when you're shopping, and think about which colors can easily adapt to the changing seasons.

Base tones are the colors that will make up the majority of your capsule wardrobe. Typically these colors offer a lot of flexibility, and the pieces in these colors are interchangable. The main tones most people base their capsule wardrobes on are neutrals:

- black
- beige
- white
- navy
- tan

Highlights and Accents

While some people are very happy with a neutral wardrobe, there are those who require a pop of color now and again. If this sounds like you, you'll want to ensure you have ample accents and highlights to coordinate with your base tone pieces.

These accent tones will serve to complement your base and can help you transition between the changing seasons. Feel free to use the color palette guide below in order to best determine the most complimentary accent colors for your wardrobe. Again, these are just a few examples. Refer to your mood board or lookbook to get an idea of what colors and color families you truly enjoy, then experiment with mixing and matching.

IF YOUR **MAIN** COLOUR IS:	CREATE EASY **COMPLIMENTARY** OUTFIT PAIRINGS WITH:	OR GO FOR A **TONAL** OUTFIT
PINK	● ● ● ○ ●	● ●
RED	● ● ● ○ ●	●
ORANGE	● ● ● ○ ●	●
BEIGE	● ● ● ○ ●	● ●
YELLOW	● ● ○ ●	
GREEN	● ● ○ ●	● ●
LIGHT BLUE	● ● ● ○ ●	● ●
DARK BLUE	● ● ● ● ○ ●	● ●
PURPLE	● ● ● ○ ●	● ●
BROWN	● ○ ●	●
GREY	● ● ● ●	○ ●

CHAPTER 6

Shopping the Right Way

Shopping can be an overwhelming experience for even the most regular of consumers and even more so for those who are searching for their first capsule wardrobe. Never fear—this step does not have to be terrifying! If you've fallen victim to the appeal of the fast fashion world, you're going to require a bit of rewiring to help you focus less on grabbing what looks appropriate and to concentrate more on building a sustainable, long-term wardrobe. To assist with that, we've compiled six helpful tips to shop the right way for the ultimate capsule-building experience.

1. Reduce Your Fashion Footprint

Earlier, we touched on how fast fashion may not be environmentally or ethically sustainable. This is something many people don't consider on a daily basis. There are actually many environmental concerns surrounding the production of inexpensive clothing, including sourcing and sustainability of the materials and fibers, and pollution created through the manufacturing process. There are also the conditions under which the garments were made. Some fast fashion manufacturers cut corners, leaving workers underpaid in potentially unsafe environments, all in the name of keeping costs low and profits high. While we can't possibly know your personal motivation factors for supporting sustainable fashion, the goal of building a capsule wardrobe allows you to support sustainable fashion, in that you are buying fewer items and you are investing in items that are made to last. Because you are being more selective, you have the opportunity to research brands and designers you feel good about supporting.

An easy way to reduce your expenses while still finding high-quality pieces for your capsule is to shop at local consignment stores or thrift shops. This is particularly useful if one of the reasons you've decided to become a capsule queen is to reduce your fashion footprint. Shopping secondhand not only makes use of items that would otherwise go to waste and end up in a landfill, it also allows you to wear fabulous clothing at a fraction of the original price. This means you can save a few bucks all while feeling confident that you're shopping mindfully and making the best use of already available resources.

For those of you thinking that secondhand clothing might already be on its last legs, here are some helpful hints to make sure you are shopping smart and not just diving into good intentions! Inspect clothes carefully, even if they still have their original retail tags. Check all of the seams for tears or stress. Some of these minor imperfections might be quickly remedied at home, but things like stretched necklines or unraveling hems might require more attention. Check for stains and pilling, as well as for signs that perhaps this garment has been very well loved. You're looking for long-term wardrobe choices, not quick fixes!

Before you plan a trip to your local mall, poke your head into some second-hand stores—they can be a treasure trove! You never know what you will find, making it an exhilarating experience that won't challenge your budget. You might find that you have a special knack for discovering amazing deals!

2. **Fashion Delivery!**

Online vintage shopping is another fantastic avenue for sustainable fashion choices that won't break the bank. With the increased popularity of online vintage boutiques, you can find flattering, unique, new-to-you pieces with just a few clicks of the mouse. It's easier than ever to locate and purchase high-quality pieces that can really make your capsule pop.

Since they're vintage, it's less likely you'll run into somebody with the same piece as you. Developing a uniform as part of your capsule wardrobe isn't about being boring and repetitive, after all! It's important to have pieces you enjoy wearing, and buying some kicky vintage pieces can further assist in developing a personal style that you can be proud of.

There are a few things to know when shopping vintage, of course. It is very important to know your measurements. While there still isn't a standard size chart, different eras sized differently, so it's crucial to look at the item's measurements to make sure it will fit. Also make sure you know everything you can about the item's condition. Just like in the consignment stores, these pieces have been worn before, plus, they may have been stored for many years.

One option to keep in mind, if you find vintage pieces you love but fear for their longevity, is to have a talented seamstress or tailor recreate the pieces using the original as a pattern. This option may not be ideal or available for every wardrobe, but it is something to keep in mind when considering the future of your capsule wardrobe.

3. **Quality, Not Quantity**

Of course there's absolutely nothing wrong with hitting the mall! As long as you are being mindful of the quality of the pieces you are selecting, you can still shop at whatever store you like.

No matter where you shop, be wary of "too good to be true" sales. They seem like great deals because they save you money, but saving money on this purchase doesn't matter if the quality of the piece is low. Do some simple math: If you purchase two shirts today for an amazing $15, and they wear out after a wash or two, it's going to cost another $15 to replace them. Rather than constantly replacing two $15 shirts, you can save money overall (and your time) by purchasing one well-made, if initially more expensive, shirt.

In addition, if you're reducing your wardrobe for ethical reasons, you'll want to do further research into the brands you love to ensure that they're meeting the same ethical considerations you have for yourself. If your high-quality jacket is made in a rundown sweatshop, do you still want to buy it? That's only one example of the questions you'll want to ask yourself as you search for your new look. The internet is full of information, so before you hit the mall, get the details on some of your favorite shops and designers. You can make a list of brands to support and brands to avoid, which will help you steer around larger shopping centers with confidence.

4. **Bring a List**

Many people find that bringing a list of the items they need to purchase to the grocery store will help keep them focused on the task at hand and away from the flashy deal signs and impulse chocolate aisle. The same concept applies to shopping for clothing, though few people actually consider it! How many times have you declared that you're "just looking"?

If you've drafted your capsule correctly, you should have a clear idea of what you're looking for when you go shopping. Don't try to buy everything at once, either. You'll only become overwhelmed by decisions and possibly rush the process by settling for whatever is nearby. That is the exact opposite of the intentional, purposeful purchases that make up a capsule wardrobe!

Take an extra few minutes to fine-tune your list before you even leave the house. Think about pieces that you really and truly need today or this week. If something that's not on the list catches your eye, make sure you fully analyze it. Does it have a place in your capsule? Does it fit on the mood board or lookbook? If possible, leave any impulse purchases behind and let yourself think about them before pulling out the credit card!

5. **Be Sales Savvy**

Be aware of your favorite brand's marketing techniques. In the digital age, many stores will offer you 10 or 15 percent off your first sale by asking you to sign up for their promotional email list, or "like" or check in on their social media page. This works out well for both you and the online shop, but be careful! Do that at every store you shop at, and suddenly, you're once more nose-diving into making more and more spontaneous purchases. **Do:** sign up for promotions from sites where you plan to shop frequently. **Don't:** sign up for a variety of promotions and use that as an excuse to shop indiscriminately!

Some shops might even have "recommend-a-friend" deals, too, benefiting both you and your capsule-building bestie. With these deals, you receive a percentage off future purchases by sending coupons to a friend. Make sure you don't justify any unnecessary purchases with the arrival of coupons and offers like this. Instead use this as an opportunity to coordinate purchases with a friend (or even a spouse).

By staying aware of new deals and sale opportunities, you can ensure you're getting the most bang for your buck no matter where you shop. Just be sure to view any promotions with an analytical eye—holding a coupon does not mean you necessarily need to use it!

6. **Focus on Fit**

As mentioned earlier, there is no such thing as a standard size. Men, women, and children around the globe typically have clothing in three to four different "sizes," depending on each brand's measurements. Since you know these numbers are more of a suggestion than a rule, do not let them impact your wardrobe.

Building your capsule is supposed to encourage you to look beyond the size tags and focus on more important things:

- Does that top fit you well and enhance your natural beauty?
- Are you comfortable wearing those jeans?
- Can you sit, stand, and move as you naturally would through the day without discomfort or the fear of a "wardrobe malfunction"?

The goal of the capsule wardrobe is to have pieces you enjoy wearing frequently. Therefore, you need to focus less on what size something is, and focus more on fit. Are the armpits too tight or too droopy? Does the waistline make you suck in your breath? Do the pant leg hems drag on the ground? These are the types of things you need to focus on, not whatever arbitrary number the designer has printed on the label.

The fashion industry has conditioned us to believe that larger sizes lower our self-worth, but considering how much your size can differ from store to store, isn't it time you stopped focusing on the number and started focusing on the fit? When shopping for your capsule wardrobe, you'll need to try on clothes, as intimidating as that may be. Try a few different sizes. Move naturally in them. Sit down. Stand up. Whatever tasks you might do throughout a normal day, try to replicate them in the fitting room area. If at any time you doubt or worry about the way a garment fits, discard that item immediately. There are too many wonderful clothes out there for you to waste space in your capsule for something you don't love!

Since the main purpose of a capsule wardrobe is to end up with fewer pieces than you began with, it's vital that the choices you make are items you see yourself in regularly. If you are hemming and hawing about whether or not a dress or a shirt looks good on you, don't commit to it. Wait for the item that fits perfectly!

Starter Pieces

Now that you're prepared to shop the right way, let's talk a bit about starter pieces—the basics of your capsule.

Your starter pieces are the go-to essentials you can see yourself wearing again and again without feeling like you've overused them. This might be your favorite tee or a pullover that you cannot live without. These items should appear frequently in your mood board or lookbook and be something you can incorporate into a variety of your selected looks.

The basics of your capsule are going to exemplify the three key factors you should keep in mind when choosing its pieces: **style, fit, and fabric.**

When you're seeking your basics, remember that these should complement your bold looks. Focus on neutral tones you can easily pair with your other items. The style of the basic pieces should be something that incorporates well with many different looks. For example, a black crew neck cotton shirt or a pair of well-fitting khaki pants might be a good choice for a basic piece, as they can be styled in a variety of ways.

The fit of your pieces, as we've discussed before, should be comfortable. Do not settle for too tight, too loose, or unflattering. Basic pieces are readily available, so do not feel you have to rush into purchasing something that is not completely comfortable. Keep in mind that you will be wearing these pieces in a variety of scenarios. Those slacks may be comfortable for sitting, but will they be comfortable after enjoying pizza and cake at a birthday party? Be prepared for many different situations, weight fluctuations, and activities.

This is where choosing high-quality fabrics will really pay off and save you some serious cash in the long run. Your basics will be what you wear the absolute most. Take your time when establishing them to avoid having to replace them frequently!

When picking your starter pieces, ask yourself:

1. Do I love this so much I'll wear it until it falls apart?

2. Does it pair well with my other pieces?

3. Can I live without it?

4. Is it made to last?

With your basics determined, you're well on your way to building your versatile capsule wardrobe.

CHAPTER 7

Care & Maintenance
to Get the Most Out
of Your Capsule

The idea of having a smaller collection of clothing may lead to you think it will be much easier to maintain your closet and your pieces. While you'll certainly have fewer pieces to worry about, there are still a few things you should keep in mind to ensure you're getting the longest possible life out of your small collection.

Wear Undershirts
Wearing undershirts will assist with keeping your shirts clean and free of pesky sweat stains. Additionally, they'll provide an opaque base layer that can help you wear more sheer wardrobe items with confidence.

The type of undershirt you choose can depend on your uniform. Men may wish to wear a light T-shirt style undergarment, while women may prefer a coordinating tank top under a blouse. You'll typically choose a neutral color that matches or fits the color palette of your outfit.

Wash Frugally

We can typically get away with wearing our clothing a couple of times before they're due for a full washing, unless we were working out hard in them or they've fallen victim to a spill. Naturally, you don't want your clothes to gain an unsavory aroma, but overwashing your clothes can make them fade, unravel, or lose their shape sooner. By washing them when it is necessary, you can extend their wearing life.

There are also a few washing techniques that are friendlier to clothing, such as spot cleaning or using cold water only, which you may wish to investigate. Many of the more eco-friendly laundry tips will enhance the lifespan of clothing, so feel free to investigate the following tips that will be both clothing and time savers in your home.

Hand Wash/Line Dry

One eco-friendly method for washing and drying clothing is to omit the high-powered washer and dryer entirely. Gently hand washing your clothes will help increase the odds that your items will keep that store-bought feel and look. Try not to line dry in direct sun, as that can cause fading. Instead, allow air circulation to wick the moisture away from freshly cleaned clothing. The kinder we are to our clothes, the longer they'll want to stick around with us.

Choose the Right Products

You can help keep your clothes intact by being mindful of the types of products you're using to clean them with. Gentle, non-toxic products made with natural ingredients can preserve the look of your clothing. For utmost care, be sure to carefully read the washing specifications labels on each item. Pieces can become damaged, bleed dye on an entire load of laundry, or shrink terribly if they are washed incorrectly.

Keep a Working Wardrobe List

We briefly touched on the idea of keeping a working wardrobe list earlier, but it's such a great way to keep your closet under control and organized. Your working wardrobe list should be regularly referenced to see which items in your capsule are beginning to show wear and tear and should be replaced or repaired. Having this readily accessible makes the upkeep of your closet easy since you know what items require urgent attention. Knowing that an item may need replacing ahead of time can prevent wardrobe malfunctions or impulse buys, as you won't find yourself bolting to the mall for a quick and easy replacement in desperation or embarrassment.

CHAPTER 8
Capsules for Every Body

It's rare to find an adult who is 100 percent in love with every inch of their body 100 percent of the time. It's important that we don't spend time feeling negative or obsessing about the body we think we want. Fashion is about making the body you *already have* feel great and look fantastic. No matter your size or shape, there are fashion choices you can make that will make you feel proud of your personal style.

It's been said that the best accessory for any outfit is confidence—and we fully agree. If you feel good wearing it, then go for it! Not everyone feels great wearing just anything, though, and that's fine, too. Make clothing choices that build your confidence, so you leave the house feeling amazing every single day. One method is to dress in clothing that accentuates the features you feel like flaunting. If you love your legs, by all means, wear some skinny pants or a skirt. If you love the way a long jacket accentuates your shoulders, throw that baby on and **Rock. Your. Day.** Intentionally dressing this way will provide a well-deserved boost to your self-esteem, so it's worthwhile to give your fit significant consideration.

Some people know right away what areas they like to accentuate, but for others, this may require some thought. We're often taught to hide flaws behind bagging clothing and thick, dark fabric, but the solution to feeling great in our clothes can start with knowing your shape. The first step in creating a fabulous look for your body is to figure out what kind of body yours most closely resembles. Stand in front of a full-length mirror, or ask a friend to help you walk through the five main body types to determine which one is yours. This is not a test. **Your body is beautiful and worthy no matter what shape it is.**

The Five Main Body Types:

Pear
Pear-shaped bodies tend to have hips that are wider than their shoulders and a small waist. People with pear-shaped bodies may have encountered the following when shopping:

- pants that fit in the waist but are too tight in the hips
- pants that fit in the hips but gap at the waist
- tops that are significantly smaller than bottoms, which can make buying a suit difficult
- tops that don't button at the bottom

Straight
Sometimes referred to as a column-body, straight-shaped bodies have shoulders that perfectly align with their hips. There tends to be little-to-no waist definition, and you may have been accused of having a flat bottom. Straight bodied people often complain that clothing makes them look too "solid" with no movement or definition. They can also have trouble filling

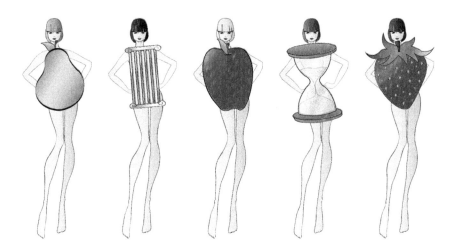

out clothing that is designed for curves, so hemlines may fall in odd places on the leg. You may have a serious collection of belts, because bottoms feel like they just hang from the waist.

Apple

Apple shapes, like straight shapes, have little-to-no waist definition, but they tend to have more rounded and wider shoulders. If you think your shoulders are broad for your frame, you are likely an apple. Common complaints from people who are apple-shaped include feeling like they're wearing a tent or a bag or having difficulty finding items that fit in the waist that don't give them a droopy bottom. Apples may have multiple pieces of shapewear in the closet as an attempt to fit into mass-produced fashion in a way that's both comfortable and flattering.

Hourglass

Hourglass-shaped people are also referred to as curvy, given their obvious waist definition, wide shoulders, and hips. If you're on the fence as to whether you're truly hourglass-shaped, try measuring your chest, waist, and hips. Typically, hourglass measurements will be similar for chest and hips and significantly smaller for the waist. Hourglass figures can be just as challenging to dress as other shapes, because the measurements often mean buying a shirt in a larger size for the chest and bottoms in a larger size for the hips. Hourglasses may feel that their small waist gets lost in excess fabric, or they might pop buttons on shirts, trying to find the right fit.

Strawberry

Strawberry-shaped bodies, also known as "inverted triangles," generally have wide shoulders, flat bottoms, narrow hips, and small waists. An easy way to find out if you fit this body type is if you generally buy larger sized clothing for your tops than you do your bottoms. Again, this shape can make it difficult to buy two-piece suits or dresses that are flattering all the way up and down the body. People with a strawberry-shaped body can also feel that they look top-heavy.

Dressing for Your Body

There are no real rules to what you can or can't wear. Like we noted before, the name of the game is confidence and comfort. However, if you're looking for some tips, some styles suit different shapes better than others. If you're looking to better define your clothing to fit your body, these general tips will guide you on the journey. Don't consider these as rigid rules, but more like ideas to get your creative juices flowing and to help lead you to clothing choices that will enhance that confidence and comfort feeling you're looking to achieve with your wardrobe.

Pear

Since your hips get attention, it's time to draw the focus to the top half of your body.

When selecting things for your top half, whether it's a jacket or a blouse, try to make sure the hemline doesn't fall at the widest part of your frame, as this will further accentuate the pear frame. Try on different lengths and styles of jackets, for example, to see what length feels the best and makes you feel less bottom heavy and more body beautiful.

Choose bootcut or straight-legged pants and A-line dresses and skirts.

If you like to wear stripes, focus on horizontal stripes for your top half and avoid them for your bottoms. Stripes lead the eyes in the direction they go, making that area seem longer than it is.

When layering, wearing heavier textures and fabrics on your top half will add a sense of density to your upper body.

Pear Clothing Tips

- Wear wide-hemmed pants, skirts, and dresses.
- Try lighter colors on the top half of your body and dark colors on the bottom for greater contrast.
- Consider cowl, boat, and square necklines to bring attention to your neck and shoulders and to add dimension to your upper body. Scarves, ties, and necklaces can also add focus to your upper body.
- Show off your shoulders with strapless dresses.
- Adding texture, such as ruffles on top, can help you balance your overall silhouette.
- Don't wear jackets that bring attention to your largest measurement.
- Have fun with patterns on top and bottom. Wild patterns and colors can be fun accents that draw the eye equally to all of your favorite areas.

Straight

Your slim hips make jeans look fabulous, so feel free to experiment with a variety of denim. People with straight-shaped bodies have the luxury of enjoying nearly every type of jeans, from skinny to full leg, as well as denim skirts and even overalls, which tend to pop up in fashion trends every few years but can be a great utility piece for running errands.

Straight lines can further accentuate your figure, leading to that "block" or "column" feeling. A-line or narrow pleated skirts and dresses typically work best to add movement and dimension to the bottom half of the body.

Avoid items that try hard to add curves. Textures (like frills) or anything bulky and wide may seem like they would give you more curves. In reality, they work against your intentions, making you feel like a wider block or column. Strategic ruffles, accents, and straps can help create shape if used in moderation (and in just the right spot), so try some options that appeal to you to see if they flatter or fall flat. Fitted shirts and tailored cotton fabrics on top will sit nicely on your frame.

Straight Clothing Tips

- Wearing long jackets may highlight your straight stature.
- Choose tops with ruffles and embellished collars to accentuate your chest.
- Add dimensions by utilizing layers, such as vests, sweaters, or jackets.
- Scoop necklines and sweetheart tops help create a wider visual perspective to the top half of your body.
- Clothing with side crunch or ruching looks great on you.

- Brightly colored bottoms will further add dimension to your shape.
- Enjoy tailored clothing to complement your natural shape.

Apple

Simple and straight lines will be an added benefit if you have an apple body type, but straight lines will be a difficult fit for curvier bodies. Tunic-style tops with well-fitted pants tend to work well if you wish to accentuate your legs while minimizing your shoulders and waistline.

A-line dresses and pencil skirts tend to be well received. Be aware of sharp angles in tailoring and tight seams that can hug curves in an unflattering way. Cardigans and swing-style coats without belts are ideal. V-necks can also be a comfortable choice that give visual dimension. Try fabrics that are thicker to streamline your look, such as tweed, silk, linen, and thicker cotton.

Apple Clothing Tips

- Monochromatic tones will look great.
- Elongate your torso with V-necklines.
- Accessorize with belts to accentuate your waistline.
- Seek out empire dress lines and tops.
- Strengthen your shape with bootcut or flared jeans.
- If you want to accentuate your legs, try shorter skirts.

Hourglass

Hourglass-shaped people may want to accentuate their curves. To do so, avoid both loose or heavy and bulky clothing as they have a tendency to emphasize the larger areas of the hourglass proportions. When selecting dresses and skirts, aim for ones with distinct waist definition, or accessorize with a belt. For pants, you're in luck, as you can get away with most bottoms, but the more defined their shape is, the better.

Hourglass Clothing Tips

- Focus on precisely tailored clothing to accentuate your curves.

- Experiment with wrap-style dresses.

- Accessorize with belts to enhance your shape.

- Wear high-waisted bottoms.

- Choose thin, light, or flowing fabrics, like jersey, rayon, or silk.

- Straight-legged pants or skinny jeans look great and can be super comfortable.

Strawberry

Strawberry shapes can be difficult to fit with clothing straight off the rack. Many people with strawberry-shaped bodies choose to tailor their outfits to accentuate their bottom halves. Well-fitted jeans and pants are a must for every body type, but strawberry-shaped bodies can be especially hard to fit. Try to purchase pants from a retailer who offers a variety of shapes and cuts, so you can find a favorite fit.

Straighter-lined clothing will help draw the eye away from the shoulders. Try fabrics and styles that hold a shape well, like waterfall jackets, corduroy pants, leather, and denim. Sleeveless tops and wrap skirts or dresses are also ways to add dimension to your lower half, while drawing attention away from your upper half. V-necklines work very well for this, too, while wider necklines can potentially emphasize wider shoulders.

Strawberry Clothing Tips

- Bright bottoms in bold patterns and flowy fabric can accentuate your legs.

- Wide-hemmed pants and skirts also add focus to the lower half of the body.

- Full skirts can balance out a body that feels top-heavy.

- Thin straps like spaghetti straps and boat necklines can emphasize wider shoulders, so choose a bottom that balances this look.

- Wear high-waisted pants, belts, and wrap styles to aid in creating a more defined waistline.

CHAPTER 9

Capsule-On-The-Go: The Quick Guide for Getaways and Trips

Another place where capsule wardrobes can excel is when you're packing for travel. Most people tend to overpack their luggage because they're worried they'll forget something important and have an "Oh no!" moment on their trip. Fortunately, that rarely happens, and the excess outfits stay folded in the bottom of your suitcase for the duration of the trip. Why not save yourself the time, energy, and luggage space by building a condensed capsule for your next getaway?

Travel Capsule Packing Tips

1. **Only pack what you truly love**

   Just as with your everyday capsule, you want to clearly define what you're naturally driven to wear in the climate you'll be traveling to. If you love the color of the bikini you bought last summer but feel uncomfortable wearing it, *don't* include it in your travel capsule. Focus on what fits you well and makes you feel good as pieces that don't will be left unworn, taking up space in your luggage that could have included pieces you would actually need.

2. Pick a color scheme

Choose a color scheme before you start packing. This way all your items are able to be included in outfits with each other. This creates a wider variety of outfit options with fewer total pieces.

3. Limit your shoes and handbags

Shoes and handbags take up a lot of precious luggage real estate. By limiting the number of each you bring with you, you'll maximize room for necessities, like a coat or fantastic souvenirs!

On most vacations, you won't require more than two pairs of shoes. That's one pair of sneakers or casual walking shoes and one pair specific to your destination. If you're heading to the beach, this might be a pair of flip-flops or sandals. But if you're heading to the mountains, a pair of sturdy boots would be more appropriate. Always pack shoes that make sense for the destination, but consider shoes that can take on double duty. For example, you might have a pair of black flats that are great for walking around but can step up and look fancy when paired with the right skirt. For men, consider suede sneaker-style shoes that can keep feet moving all day but not look out of place with a pair of nice slacks.

And handbags? You'll need enough room to carry your wallet, phone, passport, and possibly a snack. You probably don't need those two totes, a beach bag, and a backpack too. Backpacks, satchels, and hobo bags are all styles that can look great with multiple looks and serve as a handy way to tote things while shopping, when going to the beach, or on a variety of all-day adventures. Try to limit yourself to only one bag for carrying; you'll be amazed at how efficient it is.

4. **Consider chameleon pieces**

   Wardrobe chameleons are those items that can be dressed up or dressed down, being interchangeable for different situations that might arise. While we're not advocating for you to bring a chameleon piece for every "what-if" scenario, we are encouraging the consideration of pieces like long-sleeve T-shirts that can be rolled up in milder temperatures or put under sweaters for cooler nights. Think of a short jersey dress that can be dressed up with the right accessories or how a linen shirt can change appearance with the addition of a jacket and a pair of tailored slacks.

   With all that in mind, it's time to build your travel capsule.

Step 1: Basic Structure

Start with your basic structure as we did with your everyday capsule. These essential items are going to include the clothing you decide to wear in transit, so bear in mind what method of transportation you will be using. For instance, if you'll be traveling by airplane, you may want shoes that you can slip off quickly while going through security. If you're traveling by bus or car, you might want clothing that is soft, easy to layer, and won't feel too tight or constricting when spending a lot of time seated.

Your basic travel capsule for a long weekend should consist of:

✓ **Three** tops

Pack tops for a variety of occasions. Include at least one ¾"-length sleeve or long sleeve that's appropriate for the destination climate. Even when you're traveling down south where the evenings are considerably warmer than you're used to, it's good to have a lightweight long-sleeved shirt to protect you from pesky mosquitoes.

✓ **Two** bottoms

Pack a pair of formal and a pair of casual full-length pants. These items can come in handy in a variety of situations, including evening meet-ups, cool, rainy weather, or hikes in heavy foliage. Make sure you include a long-sleeved top that can pair with both of these pants, making you instantly ready for a semi-dressy or business-casual evening look.

✓ **One** sweater or jacket

✓ **One** pair of sneakers or walking shoes

If you're expecting inclement weather, some form of rain protection should be included if your jacket is not weatherproof.

Step 2: Additional Items

Once you've established your core capsule, you can start incorporating some extra items you think will be useful for your trip. These items could include a blazer, a dress, formal footwear, an additional top or sweater, a skirt, or a pair of shorts. You can also include trip-specific wear, like a bathing suit and cover-up, or clothing for specific activities, like hiking shorts, sports bras, or fleece outerwear, but remember to evaluate how often you might need these items.

Step 3: Accessorize

Finally, it's time to jazz up those outfits with stylish accessories. Since accessories are known to dramatically change the look of an outfit, it's important to include them as part of your capsule. This ensures you have the most options out of the least amount of clothing items.

A few accessories to consider including in your capsule are belts, ties, necklaces, bracelets, earrings, and tights.

Step 4: Review

Now that you've collected all your pieces for your travel capsule, you'll want to review everything you have chosen and make sure that it makes sense with your travel plans, destination, and the kind of activities you plan on participating in.

Ask yourself these questions:

1. What do I want to wear on the plane/bus/boat/car?

2. What will I wear if attending a formal dinner and party?

3. What do I need for my rock climbing/snorkeling/hiking excursion?

4. What will I wear to historical sights, museums, and shops?

5. Do I have something in case it rains or I get cold/hot?

If you've done your due diligence, you're able to pack lighter than you're used to (a bonus if you've been subjected to paying hefty fees for overweight luggage!) and you'll use everything that you bring.

Another point to consider is whether you'll be able to do some laundry while you're away. If you're able to do laundry, you'll easily extend your capsule garment wear and pack fewer overall items. You may want to pack a few additional items if your trip is longer, but keep this structure in mind. Pack for what you know you'll be doing, and try not to consider the limitless possibilities of a well-earned vacation!

CHAPTER 10

Capsules for Him

The capsule wardrobe phenomenon is a practical solution for everybody and *every body*. We have made some mention of men who have benefited from the capsule wardrobe concept, including Barack Obama, Steve Jobs, and Mark Zuckerberg.

In fact, many men tend to gravitate toward the concept of a uniform, which may make it easier for you men reading this to adopt the capsule lifestyle. You may notice that you have a dozen golf shirts in a variety of colors or that all of your work shirts are white and blue. This may mean that converting to a full capsule lifestyle is just a matter of tidying up: getting rid of pieces you never wear and building upon the looks that you already have.

But what if you decide to have a little fun? Go through the exercise of creating the mood board or lookbook. Consult pictures of your favorite celebrities, movie characters, or maybe look through a few magazines and pull some styles you really enjoy. While there's nothing wrong with the golf-shirt lifestyle, perhaps you have the opportunity to extend the wear of your pieces with a few new items.

Many men's looks have the advantage of being able to transform from casual to cocktail quickly. For example, a long-sleeved button-down collared shirt and grey slacks can look business casual on their own, but when paired with a jacket and a tie, become much more formal.

Here are some necessities that can make a variety of transitions from casual to professional:

- Basic tees: three in total, with at least two solid colors.

- Dress shirts: three in total, with one patterned (stripes, gingham, etc.).

- Casual button down shirts: two total. These can be patterned or solid

- Jeans: two pair. Try choosing two different styles to add flexibility to your wardrobe.

- Dress pants/khakis: three pair total. While having three different pairs is advisable, make sure they coordinate with all of the shirts: tees,dress shirts, and button downs. This fully extends your wardrobe.

- Blazer: Just one, unless you wear jackets to work frequently. This blazer should be able to coordinate with dress clothing and jeans to carry your look to a variety of occasions.

- Casual jacket or sweater: Again, just one. This jacket can be a variety of styles, such as a military- or cargo-style jacket, or a cardigan-style knit or wool sweater.

Accessories can include one pair of neutral dress shoes, one pair of boots, one pair of sneakers, and a belt.

If you wear ties frequently, you may have amassed a collection. As you consider your capsule wardrobe, now is a

good time to purge your collection. Any ties that are rarely or never worn can be donated, making room for ties that coordinate with your dress shirts and blazer. You can also wear ties with the casual jacket or sweater for a more professional look. One test for ties is to make sure they coordinate with all of your collared shirts. If they don't pass this test, perhaps that is not the tie for you.

Often, men have less urge to explore fashion options. Going beyond your comfort zone can actually be a boost to your capsule wardrobe and optimize the minimalist approach to fashion. There are many pieces that can be incorporated into a men's capsule wardrobe that may have previously been "wishful thinking" pieces. Here are some examples of pieces that may seem extraordinary but can actually bolster an everyday wardrobe:

- The leather jacket or wool coat. Outerwear is a must in many climates, so why not have fun with a piece that is fashion fluid? Picture a leather jacket with a T-shirt and jeans. Then put that same leather jacket over a crisp white Oxford shirt and a solid black tie. Try the exercise again with a longer wool coat. Notice how the overall look changes?

- Henley tops. Worn by themselves, they have a rugged, woodsy appeal. Worn under a blazer, you have an artsy, hip expression.

- Flannel or plaid print shirts. Another type of shirt that can go from the great outdoors to martinis, bolder print shirts in a casual fabric are very versatile.

- Sweater vests. This layer can be thrown over a plaid shirt to dress it up or layered over an Oxford and tie to add a slightly more casual feel to the outfit.

Suits are another area where the uninitiated can panic. Obama has the right idea: blue and black are classics that can work in nearly every situation. Grey is another shade that won't stick out in a majority of situations. For many men, one suit will fill your needs for every formal event in the foreseeable future, so make sure you pick one that you feel comfortable wearing nearly anywhere.

The most important feature of a suit is the fit. While there are constantly changing fads in lapel width, trouser break, fabric, patterns, and shoulder width, a well-fitting solid-color suit in either wool or cotton can be en vogue for years to come. Make sure the seam of the shoulder meets the actual spot where your arm and shoulder meet. The jacket should fall straight to where your knuckles rest naturally and button with room to place a fist between yourself and your jacket. The sleeves should be no longer than your thumb. The cuffs of the pant legs should rest at the top of your shoes.

While picking out a capsule may seem like an easy chore of buying an armful of the same exact item in a variety of colors, this can be your chance to really extend your wardrobe's capabilities. Always take into consideration how you might actually wear your items; for example, if you haven't worn a tie in over twenty years, you can skip buying new ties. As we practiced in the drafting of the capsule, it's all about being realistic and finding looks that make you feel comfortable and confident!

CHAPTER 11

Capsules for Work and Capsules for Play

Most people do not live a static lifestyle—they may dress a certain way for work, but activities, including sports, exercise, and hobbies, may require a different set of clothing altogether. Now that you have the right idea for building a capsule wardrobe, consider how many capsules you need.

That's right! You can have more than one capsule if need be. Some people have very distinct and small capsules. Rather than one thirty-piece capsule, they may have a fifteen-piece work-clothing capsule, a ten-piece casual capsule, and a six-piece capsule of workout attire.

Workout clothes, for example, are going to require frequent washing due to their purpose. If you participate in intense physical activities a few times a week, you may wish to create a workout capsule.

This could include:

- three pairs of leggings
- two pairs of shorts
- five tank tops or athletic tees (or combination)
- five pairs of athletic socks
- two sports bras
- a hoodie or cool-down top
- a well-fitting pair of workout shoes

Some hobbies, like gardening or painting, may also require their own uniform. While clothing that has been retired from the regular rotation capsule is a good go-to choice for adult "play clothes," you may also find that adding items like cargo pants, with many pockets to store some of your needs, might be helpful. You may also invest in a pair of bib overalls or an apron to protect your clothing or weatherproof jackets and footwear.

While there are too many activities to provide a comprehensive list of every possible capsule combination, always apply the capsule theories to your fun clothes. For example, choose pieces with purpose. Do all of your tops coordinate with all of your bottoms, allowing you a fully flexible wardrobe with multiple possibilities? Also, don't impulse shop for activewear. While that pair of wind pants may seem very cool, are you going to enter them into your capsule or wear them for one cold-weather 5K and forget about them? Make sure all of your fun clothes fit just as well as your regular capsule, especially if your activity includes a lot of movement. Seams that chafe or poorly constructed clothing that unravels is especially frustrating when you're on the go!

CHAPTER 12

A Capsule for Every Situation

While some people prefer to divide up their capsule into things like "A Work-Ready Capsule" and an "Everyday Casual Capsule," or a "Spring Capsule" and a "Summer Capsule," it can quickly get out of hand and disorganized. This can diminish the functionality of your capsule and encourage you to buy new pieces.

With the exception of specific-use capsules, as mentioned in the previous chapter, you should have a wide variety of pieces that can be utilized in many settings and seasons.

Take a look at the pieces you have identified on your mood board or lookbook or that you have currently in your closet. Keeping in mind the flexibility and fluidity of a capsule wardrobe, make a note of what pieces can be used in more than one season. For example, a short-sleeved shirt may be a stand-alone top in the warmer months but pair well under a wool blazer in the cooler months. This extra bit of attention will be instrumental in keeping your wardrobe small while maximizing its potential for usability.

We'll go through some possible transition scenarios to help you consider how your wardrobe items can do double-duty. If you're more of an experiential learner, feel free to pull out the mood board, lookbook, or even start pulling items out of your closet!

Transition 1: From the Office to the Party

The capsule you have built so far should keep in mind any office dress code or requirements. Your capsule will not work if you do not build it around your daily essentials!

In this example, we'll pretend we work in a hypothetical office with a business casual dress code. After work, we have a party to attend at a trendy, casual restaurant.

Layers can be a great friend when it comes to any type of transition. Let's say you walk into work wearing a bright short-sleeved shirt with a blazer over it. On the bottom, try one of your coordinating skirts or dark slacks. To quickly turn this into a more fun outfit, ditch the blazer. If it's chilly, add a fun jacket, like denim, or a light sweater. This is the "swap" method, in which layers are easily replaced with other similar items. You can also swap your formal work shoes for something bright and comfy or trade work pants for jeans. Swapping out those heels for a neutral flat-footed shoe makes it possible for you to dance the night away.

One of the easiest ways to transform your work attire into a more relaxed look for evening is to utilize accessories. As we mentioned before, accessories have the ability to make any outfit look completely different by adding or taking away pieces. Bring a complimentary necklace and a pair of matching earrings to kick your appearance up a notch after you've left the office. Scarves can also be an easy complement. For men, ditch the tie, and consider wearing a T-shirt with a blazer for a contemporary yet casual after-work look. Remember—these are all the mix-and-match pieces you already have!

Transition 2: Work from Home to Formal Networking Event

Many people assume that working from home is a commitment to pajama pants and T-shirts, but that's not always the case. Those who work from home may have the benefit of no formal dress code, but they still have things to do and places to be. In this example, let's say you work from home, but you have a Skype video meeting at 3 p.m., after which you have to pick up the kids from soccer practice and then somehow be at the fancy hotel downtown for a formal networking event by 7 p.m. This transition will only work with the fewest possible costume changes.

For a daytime look that will be appropriate on that Skype meeting and require only minor updates to look formal just a few hours later, start simple and let layers and accessories do the heavy lifting. Start with a plain, dark T-shirt and jeans. This will be super comfortable for working from home. To make this basic ensemble suitable for errands you need to run, including picking up the kids, grab a bright or patterned blazer. Once you return home, swap the tee and jeans for a dress and keep the blazer. For men, ditch the jeans for dress pants, and add a long-sleeved dress shirt over the tee but under the blazer. As always, accessories will help bring everything together.

Transitioning from Season to Season

Another type of transition that involves your wardrobe is seasonal transitions. Many of us live in areas where the seasons change gradually, so you may need a variety of layers all year round but have a massive "summer" wardrobe and "winter" wardrobe that are stored in the off-season. Let's do a simple exercise to demonstrate how your capsule pieces can make rotating clothing for the seasons a thing of the past, adding simplicity to your closet!

Let's start by assuming you built your first capsule wardrobe in the spring.

Below, we'll go through the four seasons and suggest ways to seamlessly morph your capsule into a functional summer wardrobe, then cozy up for fall as the temperature drops, and finally bundling up with the dark, cold days of winter.

Of course, this is only to serve as a guideline to help you rummage through your own styles and arrange your capsule wardrobe accordingly.

Spring into Summer

First, let's look at what you have included in your spring capsule.

1. **Basic T-shirt**

   Everybody needs a high-quality T-shirt as part of their spring capsule. Those warmer days tend to sneak up on us and you'll want something that is functional, with and without a sweater. Choose a style (V-neck, scoop, or crew) and stick to it. Consider solid colors or neutrals for optimal versatility.

2. **Rain Boots**

   Spring means showers (and as the song goes, then come the flowers). If your region is known to get a sprinkle, invest in a good pair of rain boots or weather-proof footwear. Water-resistant shoes or boots are an essential item to every capsule. Consider how frequently you can wear these boots or shoes too. While rubber boots in a wild pattern can be fun, a weather-proof leather boot can be worn with jeans, leggings, and in some cases, even make the transition into the workplace.

3. White Jeans

Spring is the season of new beginnings and a festival of colors. What better way to celebrate new beginnings and offset those fun spring colors than with a crisp pair of white jeans? White is fabulous because it pairs well with every color—neutrals to pastels and neons, and everything in between. Since we're in spring, look for a versatile pair of jeans that aren't stifling in the warming weather but aren't cropped or sheer.

4. Denim Jacket

Aside from being a great way to keep the wind away in the spring, denim jackets are very versatile pieces you can include in your wardrobe, making it an essential to your capsule. Denim can be used to make a floral dress look casual, or it can serve to add some edge to a casual look. It happens to work best with the one spring essential: the basic T-shirt. The right jean jacket can take any outfit from dressy to chill and comes in a variety of shapes and shades that can be incorporated in nearly every capsule.

5. Trench Coat

Trench coats are one of the most useful pieces you can include in your capsule because they can be utilized from spring through fall. Include a neutral-colored trench in your spring capsule you can see yourself wearing in other seasons, even if that bright red trench from Saks is calling your name. You may wish to choose a coat that can easily slip over everything from T-shirts to light sweaters.

When Summer Turns to Fall

As we drift from summer to fall, shorts, cropped pants, and skirts may leave us feeling chilly, while heavy sweaters are too much. When transitioning from your summer capsule, bring in lighter sweaters and long pants to help fight off the chill as the weather begins to change.

1. **Basic T-shirt**

   You'll quickly understand the value of incorporating a few basic T-shirts into your capsule wardrobe, given how functional they are. This is a good time to inspect your capsule T-shirts to see if the basics you've worn through spring and summer need mending or replacing.

2. **Tank Top**

   The weather is cooling off, but that doesn't mean it's time to put away all those tank tops. Keep a few of these on hand through the fall to put under other shirts on chillier days or to layer underneath a jacket or cardigan for a fast and easy look that's not too hot or bulky.

3. **Knee-Length Dress**

   A tank-style dress, for example, can be used as frequently in the fall as it is in the summer, if you prefer a cute dress in the warmer months. Adding a denim jacket to a knee-length dress is a great versatile look in the cooler days, but you can also wear a blazer or nice sweater over these dresses to make them work-ready. By ensuring you have at least one knee-length dress, you're preparing yourself to be comfortable both in the blistering heat and on the cool, windy days.

4. **Lightweight Jacket**

A suede jacket or blazer can make a casual look pop in a crowd or give an earthy vibe to a work look. They are super useful during the transition into fall as it becomes cooler outside. They are versatile, changing a look in a matter of seconds and on the go. Jeans and a turtleneck can be a great look for men and women, but a lightweight suede jacket can take that look up several notches!

Layering Tip:

As the weather starts to turn, start implementing layering techniques. Layering is the key to making your outfits transition in-between seasons. Things like cardigans, waterfall jackets, denim, flannel, and suede can be thrown over basic tees, long-sleeved shirts, and dresses to create a variety of looks. Ideally, you want these items to pair well with all your styles and colors.

Fall Essentials You'll Use in the Winter

Moving from fall to winter is one of the least complicated changes your capsule will endure, especially if you are already equipped.

1. **Long-Sleeved T-shirts**

It's good to have long-sleeved T-shirts on hand, because they work as a base layer or a stand-alone top. They're a great go-to to throw on when you're heading out to run some errands or hosting company at home. A solid-colored long-sleeved tee can pair with jeans, khakis, skirts, and more.

2. **Sweater**

For those days when a long-sleeved T-shirt isn't cutting it, a nice sweater can make your look formal when paired with a fine pair of slacks. Or combine it with some loose-fitting jeans for a laid-back look.

3. **Jeans**

By now, you are probably aware of how useful it is to have a well-fitting pair of jeans on hand. You'll use them in every season. Skinny jeans can be worn with boots, and comfy jeans can be paired with a slouchy sweater for a casual, comfy look.

4. **The LBD (Little Black Dress) and the Black Suit**

As we get to Thanksgiving and Christmas, you'll be invited to social events and dinner parties. A clean, fitted black dress or well-tailored black suit will be worth investing in if you don't already own one.

Winter Melts into Spring

We're coming full circle now. You'll notice that your spring capsule included the largest number of options. That's because most of those pieces are being folded into the following seasons as the days progress.

1. **Long-Sleeved T-shirt**

You'll certainly be grateful for your long-sleeved T-shirts as you're cozying up to the fireplace, sipping your hot cocoa, and watching your favorite winter classics.

2. **Cardigans**

A cardigan sweater is a fantastic layering tool (and you probably already have one from your fall capsule collection). Go for quality here, as cardigans can easily be implemented into any look. You'll have difficulty keeping your hands away from it through the winter.

3. Jeans

Jeans are at the top of my list for capsule pieces that are always necessary. Since it's used so frequently, sticking with a solid color like black or blue is ideal, if you don't want to invest in multiple pairs.

4. Boots

If you live in a snowy region, winter boots are a MUST. And as that snow starts to melt in the springtime, you'll still be using them, making them an important essential for your winter-spring transition. Warmer regions don't need heavy duty winter boots lined with fur. Ankle boots (cowboy boots if you're feeling a little funky) suffice in these climates and can complement any look year round.

There are no hard rules when it comes to choosing the pieces you can utilize year-round. It's important to take some time to consider what those pieces might be before you fine-tune your capsule. The extra legwork in completing an exercise like this one will help keep your closet organized while making transitioning between seasons a breeze.

CHAPTER 13

Capsules for Kids of All Ages

Capsule wardrobes aren't just for adults.In fact, the concept also works for children of all ages. Now you may be thinking that couldn't possibly work, since children often require multiple wardrobe changes a day, due to mess, activities, and more. Plus it seems like children grow out of their clothes every day! With some strategic planning, you can really get a handle on your children's cluttered closet space by adapting them to the capsule lifestyle. Just like adults, there may be multiple capsules required, like "Play Clothes" and "School Clothes," but overall, it's very easy to implement capsules with kids as they do not require the same range of day-to-day transitions as adults. For the most part, children from birth to the teenage years can get away with one fun outfit all day. It's well worth the time to live minimally.

Baby capsules should include a variety of short-sleeved onesies, long-sleeved onesies, and pajama or pants outfits. It may feel like a baby capsule is more extensive than an adult capsule, but there are good reasons for that. Little bodies aren't able to regulate temperature as well, so babies can get hot or cold quickly. Furthermore, diaper blowouts are an unfortunate and messy reality! While it can be tempting to buy every little adorable outfit for a new baby, try to stick to necessities. When compiling a baby capsule, consider these basics:

- 3-5 undershirts/base layers
- 4-6 sleepers
- 8-10 onesies/bodysuits (long + short sleeved)
- 4-6 pairs of pants/leggings
- 1-2 rompers/overalls
- 1-2 layering pieces (sweater or hoodie)
- 1 cap/bonnet
- 6-8 pairs of socks

The same rules apply as the adult capsule: go with neutral colors that mix and match well. That way, when Sleeper #2 succumbs to an after-lunch urp accident, you aren't scrambling to find a pair of pants that can coordinate with the remaining pieces. Sticking with the capsule format for babies will extend clean clothes so much further and prevent you from buying all of those super-cute but unnecessary items that your baby might grow out of before they get the chance to wear them. Also consider forming capsule groups with other parents in the area. As babies grow out of their clothing, gently-used items can be swapped and reused!

Capsules for children ages 3-10 should ideally have between twelve and fourteen items total. Their entire wardrobe can fit in a single dresser drawer. How's that for keeping tidy! Seasonally, that breaks down to having about six tops and five bottoms for school and nonplay wear and a dress or a suit for dress-up occasions. As with adult capsules, you will get the most flexibility out of pieces that can coordinate easily with each other. Since children in this age range are exploring with dressing themselves, this mix-and-match lifestyle can make getting dressed a breeze for both tots and parents.

Once children start heading into their teenage years, they may have some very distinct and definite ideas about what they want to wear. That's part of growing up and gaining independence, so it should be fully expected. If your teen or tween is interested in continuing the capsule wardrobe format, show them how to make their own mood board or lookbook. They may find it very enlightening to look through their fashion choices—many teenagers create a uniform without even consciously considering it!

What Are the Benefits to Kid Capsules?

Much like our own capsules, there are many benefits to adopting the capsule lifestyle with your kids. With how quickly our kids grow out of their clothing, building a capsule is one of the most cost-effective decisions you can make. By carefully selecting your kids' clothing pieces to be versatile and long lasting, you'll save money over time, which can be used for other things like your kids' college funds.

You'll also be getting the most out of your purchases because instead of your kids constantly reaching for the same shirt or pair of pants, they'll be rotating their items on a regular basis.

If you've ever thought that you could use a few more minutes in the morning to get yourself and your kids ready for the day, shave off some pesky decision-making time and build them a capsule that'll ensure they look and feel their best every single morning.

The Six Key Questions to Ask Yourself When Building Your Child's Capsule

1. Is it simple?
2. Is it comfortable?
3. Is it built to last?
4. Is it flexible and versatile?
5. Is it affordable?
6. How quickly will they grow out of it?

Building Your Child's Capsule

Pick a day where your kid is out of the house and go through the closet. If you try to eliminate items from their existing wardrobe while they are there, chances are good that they'll want to keep everything even if it doesn't fit anymore.

First, take all of your child's clothes and make sure they've been thoroughly washed, unless they have never been worn. If the tags are still on them, fold them neatly and lay them on top of the pile for first picks. Unworn clothing is easily resold, especially children's clothing. After folding the washed, gently worn clothes, lay them out in front of you so you can see what you have and start connecting what your child will like to see in their capsule. If you're noticing a lot of blues, it's safe to say you should build your capsule around that hue; if you change your child's style drastically, there's a chance they won't like it and tantrums will ensue.

Make a collection of all the key items your child cannot do without as they tend to hold onto many sentimental items. Some of these items can find a new home in the "Play Clothes" capsule. If you have a plethora of these items, then you can consult your child to decide which items are the most important. Until you're sure those items won't be missed, place those lesser-worn items in a box for safekeeping.

Making the Most Out of Your Pieces

As you know now, the more versatile the capsule is, the better it serves you. Below are some helpful tips when choosing your colors and essentials.

1. **Base Color**

    As mentioned earlier, if your child already has a love
    of a specific color, you'll want to tie that into your
    child's capsule and use it as a base. If your child is too
    young to know any different, then pull color palettes
    until you find one you like. As with your own capsule,
    it's important to consider what colors work throughout
    the seasons to make shopping for replacements easier.
    For children, you may want to steer away from light
    colors that can easily get stained.

    With that in mind, choose a base color for your child's
    capsule. Black, gray, navy, or brown tend to be good
    base colors for both boys and girls, and these colors
    transition beautifully through all the seasons.

2. **Accent Colors**

    Once you've decided on a neutral base, choose two colors
    as accents. If you chose a navy base, light gray and black
    pair well. Or, if your child can't let go of a favorite bright
    color, such as pink, pair it with blue and navy.

3. **Patterns**

    Choose three patterns you'd like to incorporate into
    their capsule theme. If you don't have an affinity for
    any patterns then stripes, florals, and polka dots are
    commonly recommended staples.

Kid-friendly Essentials

Now that you've determined your theme and colors, you're
probably wondering how to break up your capsule and what

essentials to have on hand. This will vary depending on where you live, so bear in mind any regional needs.

Tops

Have twice as many tops as you do bottoms. Not only will you want long- and short-sleeved tees, but you'll also want some button down shirts that can work at school or out and about. This will enable you to layer pieces, which are highly practical in the transitioning seasons, and it adds flare to your child's style.

Shoes

If your child is over two, you're going to want to make sure you have four pairs of shoes. These should be:

- ✓ a pair of sneakers for playing and running
- ✓ a pair of weather-proof boots in case of inclement weather
- ✓ a pair of dress shoes for more formal occasions and outings
- ✓ a pair of season-specific shoes, like sandals in the summer or nonslip boots in the winter.

CHAPTER 14

Capsules Over Fifty

For the fashionista over fifty, you may feel like you're growing out of your younger styles. Your body is undergoing changes, so this is a great time to reevaluate your capsule wardrobe and build something that's comfortable for you, while still being stylish and modern.

For this, we're going to walk you through an example of a capsule wardrobe that would complement any woman over fifty.

Colors

It's time to simplify. Neutral colors like black with accent tones work well in all seasons. Your wardrobe will remain stylish for years.

Stick to solid colors for your capsule bases. If you want to add prints, have fun! But still, don't go overboard, and consider your overall capsule. If you want to buy a bright pink fringe leather jacket, go for it—as long as you have the solid tops and the right pants to carry it off a variety of ways. Less bold individuals may wish to consider low-key patterns, like stripes, and colors that coordinate easily with all of their pieces.Remember: the goal is to feel good about what you wear at any size, any shape, and any age.

Accessories

When shopping for accessories, focus on high-quality, versatile pieces and bigger brands for longevity and style. The goal is to make it easy to pull a piece that is just the right touch for your chosen outfit.

For those with issues like arthritis, consider necklaces that can be slipped over the head or bracelets that don't have a tricky clasp. Bigger statement jewelry can be fun for the bold middle-aged woman and give you a playful accent to your already stylish wardrobe.

Choose a handbag with a neutral silhouette as this will help define your style as it encourages your outfits to look modern and hip. Consider a smaller handbag or even a cross-body bag, which will make accessing your keys, wallet, and other essentials much easier than a bulkier bag.

Shoes

Find shoes that are comfortable and stylish, enhancing your overall look. Your capsule wardrobe should consist of five pairs of shoes to transition in between events and for daily wear.

Low wedges work well with any skirt or pants for casual days, while mid-heels dress up your outfit for formal and special occasions.

Sneakers are always good to have on hand for maximum comfort and versatility. They can be worn with pants, jeans, and even dresses for errands or road trips while still looking chic.

Keeping a pair of flats in your arsenal is wise for all ages. You can use these as a bonus accessory by wearing a bolder and patterned flat to express your individuality.

With sandals, black and metallic looks are always in style and can match a variety of outfits.

Bottoms

Choose pants that don't flare out too boldly and are made with a comfortable fabric. Ponte trousers, palazzo pants, stretch jeans, and classically cut jeans won't dig into your midsection and are always current. Many of us discover that your body shape and weight changes as you age. This is perfectly natural! Just be sure that any clothing you choose continues to make you feel confident and comfortable!

When looking at capris, stick to ones that crop closer to the ankle than your calf. Mid-calf capris often make legs look shorter. Make sure the cut is slim or straight, a modern and purposeful look.

Dresses

Dresses are highly versatile and essential to every woman's capsule. Easy-to-wear dresses like jersey shift dresses can be dressed up or down and have no ill-placed buttons or zippers to deal with.

They can be paired with heels or strappy boots for evenings, ballet flats for an afternoon at the office, or with a solid pair of sneakers for errands and general walking.

Stick to neutrals like a black or a grey as it will give you the most options for accessorizing and layering (you know, like that bright pink leather jacket!).

Layering

With age, some people find themselves more sensitive to heat and cold, which makes carrying a light sweater inevitable.

Blazers or short jackets are useful additions to any wardrobe and can turn most bottoms into a suit when paired strategically. Lightweight stretch fabrics will flatter your curves and help you feel comfy no matter what the temperature. Layering is more useful now than ever as it's not only used to transition the seasons, but to enhance your figure.

T-shirts

Stick to softer lightweight fabrics like silk or polyester as they drape over your figure nicely without adding bulk to your shape. Try to pick T-shirts with rounded necklines as they will add length to complement your neck.

As with capsules for every age, the following rules remain true:

1. Focus on **Essentials First**
2. **Quality,** Not Quantity
3. Dress for **Your Body Type**

CHAPTER 15

Capsule Accessories

Some people shy away from accessories when they're building their capsule wardrobe because they think they contribute to the clutter we're trying to avoid. However, accessories are key components to every capsule. Throughout the preceding chapters, we've touched on a few examples of how accessories can be incorporated into many different looks, but if you want to clearly define your accessory essentials, this is the place to be.

Accessories are vital because they're the key to making our minimal wardrobe as versatile as possible. To demonstrate this principle, put on a T-shirt and jeans. In front of a mirror, add a necklace. Then swap the necklace for a scarf. It's three different looks!

Just as with your clothing, you want to focus on accessories that serve more than one function: they can be paired with anything, they work across seasons, and they are not fast fashion. These are pieces that you are going to be keeping for a long time, so consider your accessories as investments, rather than impulse purchases.

Feeling skeptical? Here are reasons why you shouldn't dismiss accessories when building your capsule:

1. They're one-size-fits-all. You don't have to worry about growing out of them and they're perfect for sharing and trading with friends when you want to spice things up even further.

2. They can be reused indefinitely and paired with anything. Your long necklaces can look great on a dress for a dinner party *and* pair with your blouse for a more laid-back look.

3. They require little maintenance aside from an occasional polish or wipe.

4. They're small and easy to stow away, eliminating the clutter in your jewelry box.

5. They contribute to your unique style, enhancing that confidence you're working to foster.

Accessory Essentials You'll Want to Have

When it comes to purses, you only need three—each one serving a unique function. For instance:

1. A clutch or wristlet for when you just need to grab your wallet and go.

2. An over-the-shoulder/satchel for casual outings.

3. An everyday handbag, like a large tote that fits all your essential items like your wallet, keys, laptop, or books.

With jewelry, begin with pieces you can use daily, such as a small pendant or chain. From there, you can then add statement pieces that really convey your personality. Bold colors make neutrals pop and change the look of an outfit. Silvers and golds are classic, timeless, and can work with anything. Vintage jewelry and pieces that you find on your travels can add so much dimension and personality to even the most basic outfits. But remember: this is not your chance to rabidly purchase every piece you see! Instead think of at least three outfits that can work with that piece before you make the investment.

Scarves come in a variety of fabrics and weights. A lightweight silk scarf is great for warmer days, making your outfit pop. During the cooler months, a thicker wool scarf is ideal to keep warm and to offer protection from harsh wintry winds.

Hats for both men and women should be functional, first. If you live in a colder, snowy region, you'll need to make sure you have adequate gear for the climate. Wool caps, felt fedoras, and corduroy newsboy caps can keep the noggin warm when harsh winds blow. In warmer climates, you'll look for protection from the sun with breathability: straw hats with wide brims, baseball caps, and visors.

Sneakers, Wedges, and Flats

You're likely to find yourself spending time outside throughout the year, going on casual walks with friends and enjoying outdoor events that range from super informal to very dressy. For that reason, your capsule should always include a pair of good walking shoes or sneakers, a pair of wedges or dress oxfords for dressier outings, and flats for general use. Flats are an exceptionally flexible type of footwear that can be worn from spring through fall and can be worn with everything from dresses and skirts, to skinny jeans and leggings. Flats look great in any environment and are comfy enough to allow you to stand and walk for hours at a time.

Men will also want comfortable walking footwear, and as the weather clears up, it's time to pull out your leather shoes and put away the thick winter footwear.

Sunglasses

Many people associate the sun with the summer months, but UV rays and sunshine can cause eye damage all year round. Be prepared to protect your eyes with a pair of sunglasses you love and that coordinate with your capsule. Timeless looks like aviator glasses and cat-eye styles can work with outfits that are casual or dressy. Choose a frame that looks flawless on you; after all, our faces stay the same.

CHAPTER 16

Common Mistakes Shoppers Make

Building your first capsule wardrobe, although easy in concept, is often challenging. To help you avoid some of the pitfalls many people face when taking the plunge into their first capsule, we're going to share some **dos** and **don'ts**, so you can be prepared for an easy transition into your new minimalist capsule lifestyle.

The Dos of Capsule Wardrobes:

1. **Planning is Key**

   The prospect of starting your first capsule wardrobe is exciting and often leads many new capsule users into a frenzy. While it's great to fantasize about the kind of clothes you'd love to wear and completely transform yourself while building your capsule, it's vital that you plan ahead for the practicality of your lifestyle. If you envision yourself as someone who wears suits and blazers to work but, in reality, work from home in casual pants and a T-shirt, you need to pause and plan for reality. It's important to be honest with what you will wear before committing to your capsule plan. Remember, capsules are supposed to make these kinds of decisions easier, not harder.

2. Develop Your Personal Style

Your capsule is a great opportunity to further develop your personal style. Choose pieces to reflect your current lifestyle and experiment with your style. Search on Pinterest or through magazines and catalogues for additional inspiration, and think about what qualities you enjoy in clothing. Remember, you want to feel comfortable and confident at all times!

3. Focus on Quality Pieces

If you don't focus on quality when you're building your capsule, you'll end up with worn-out outfits that are falling apart at the seams. This is your chance to get sustainable, long-lasting pieces that can carry you from season to season. You'll be wearing these items more often than usual, so invest in pieces that are built to last.

4. Organize Your Wardrobe

Regardless of how many items your capsule consists of, if you don't stay on top of your organization, it's going to be difficult and stressful to pick out an outfit every day. One of the main advantages of building a capsule wardrobe is how easy it is to maintain; don't slack.

5. Purge Intelligently

While you may want to simply trash everything you have and start over, that's not the best approach. Instead, divide your existing wardrobe into piles of clothing that can be transitional, clothing that you never wear, and a pile of solid maybes. If you never wear it, sell or donate it immediately. For the maybes, take the time to try pieces on. If you don't like the way they fit, add them to the "sell or donate" pile. If you can't quickly find another piece in your collection to complement that item, add to the "sell or donate" pile. If you're just not sure, it's OK to store these clothes while you build your capsule, but a good rule of thumb is that if you haven't worn it in a month, it can probably move on.

The Don'ts of Capsule Wardrobes:

1. Obsess Over Numbers

One of the major reasons to have a capsule wardrobe is to make the most out of a minimal amount of clothing; you'll have to consider what will work for you. A thirty-six--piece capsule might fall just shy of your needs to be functional. It's important to recognize what will work for you and amend your capsule accordingly. It's okay to review and reassess your capsule as you go; just be mindful of the choices you're making and why. Also keep in mind—you can create more than one capsule!

2. Copy Someone Else's Style or Follow Trends

It's okay to use somebody else's capsule for inspiration, but copying it outright is going to do you more harm than good. When you copy someone else or impulse buy to keep with current trends, you're not properly looking at what is functional and effective for you and your capsule. Someone who wears dresses daily is going to have a significantly different idea of a functional capsule than you might if you're always in athletic gear. Save yourself a headache and review other capsules for inspiration only, not guidelines.

3. Be Too Rigid

Speaking of inspiration, this guide is just that: some ideas and tips for getting started with the capsule wardrobe lifestyle. It is okay to find that some rules or suggested pieces outlined don't apply to you. Your wardrobe shouldn't be surrounded by rules; it should be as flexible as it is organized. Everyone is starting from a different place when it comes to finding a way to organize their clothing and their lifestyle, and we can't possibly guess at everyone's needs! The goal of a capsule wardrobe is to simplify your wardrobe down to pieces that make you feel confident and comfortable and can be worn in a variety of scenarios. Want to wear a summer dress in the winter or feel like replacing your favorite T-shirt earlier than planned? The choice is yours!

4. Play It Too Safe

While building your capsule out of solid neutral colors is a great way to get yourself started, it can get dull if you play it too safe. Be sure to include a couple of fun pieces to kick your wardrobe up a notch and save yourself from getting bored. Bright accessories are great for this.

Your style may change from spring to spring, and that's fine. Don't feel like you have to wear the same capsule. The beauty of a capsule is that the core pieces should be easily incorporated into any changes in your style, lifestyle, or location.

Essential Shopping Tips

Inevitably, you will have to replace your pieces after they wear out. When you do, follow these three tips:

- Look for clothes that follow your current style.
- Don't buy something that's the same as your other pieces.
- Make sure you love every purchase before you commit.

CHAPTER 17

The 21-Day Capsule Challenge

Now that you've read through all our key tips for building a capsule wardrobe, you're practically an expert. You may also feel more nervous than ever about taking the plunge into a new lifestyle!

The 21-day capsule challenge was designed to ease the transition into capsule wardrobes. If you feel ready to take on the challenge, delve into the fabulous world of sustainable fashion, and adopt a minimalist lifestyle—it's time to dive in!

The 21-Day Capsule Wardrobe Challenge

1. Go through your wardrobe and select twenty pieces you believe will sustain you for the duration of the challenge. Since you're just getting your feet wet, these items don't include your accessories, pajamas, any workout gear, or heavy outerwear, if it's cool in your region. Just select tops and bottoms.

21-Day Challenge Example:

- 2 T-shirts (white and black or stripes)
- 1 cardigan (black)
- 4 dress shirts (white, blue, plaid, and polka dot)
- 1 blazer (red)
- 1 leather or fun jacket (black)
- 3 jeans (including skinny jeans, jeggings, or colored denim)
- 3 pants (corduroy, black, khakis)
- 2 skirts
- 1 dress or suit
- 1 pair of flats or dress shoes
- 1 pair of boots

2. Remember to try to choose pieces that are versatile and functional for your lifestyle. You will be living in them for the next 21 days after all.

3. Clean and organize your closet so that all your unused pieces are out of sight and out of mind.

4. For the next three weeks, alternate and mix and match your twenty pieces to create as many unique styles as you can. Document them, take photos, and make notes of pieces that you find more versatile than others. Consider the usefulness of layering pieces, particularly if you are transitioning into a new season.

5. Use these three weeks to experiment with how far you can stretch those twenty pieces into new outfits. Play with accessories and try to avoid buying anything new. You can use this time to save money to buy capsule pieces after the three weeks.

6. In the final week of your challenge, start thinking about how you can build your next capsule. This will teach you the importance of having a working wardrobe and to look ahead in the future for transitional periods and the needs of the next season.

7. Enjoy the challenge! While you may find that twenty pieces is more stressful than your current wardrobe setup, try to have fun with it and come up with as many styles as you can. You might surprise yourself. This is only three weeks out of your life, but it could save you time and money over your lifetime.

CHAPTER 18

The Future of Fashion

Are Capsule Wardrobes Timeless?

Capsule wardrobes are not a new phenomenon in the slightest. Soon we'll be celebrating fifty years since the introduction of this life-changing concept. In that time, we have seen fashion go through many changes, like the raising of hemlines on dresses and skirts and the lowering of waistlines, the popularity of ethnic prints and psychedelic patterns on tops, and the introduction of pumps.

More recently, capsule wardrobes have seen the fashion world develop into fast fashion, including the degradation in quality of materials as manufacturers quickly started to churn out new styles and market them to the world in rapid-fire succession. The influx of fast fashion has made it challenging to find high-quality pieces that are as timeless as they are modern and have significantly increased stress and anxiety in consumers everywhere. All the while, capsule wardrobes have remained quietly in the background and adapted to global trends like shifting into more eco-conscious solutions to everyday wear.

With the advancement of social media and the dependability of the internet, capsule wardrobes can now take center stage. Socially conscious fashion followers can build their own sustainable capsules in unique and creative ways while sharing with the world the amazing benefits and positive impacts capsules have had on their lives. The capsule wardrobe movement is on the rise, given the advantages to adopting a minimalist lifestyle in a constantly changing world. Capsule wardrobes have allowed us to take back our lives from a deluge of decisions and develop a system that works for our daily needs.

With so much to keep up with in our "always-on" society, choosing what to wear should take up as little time as possible, so we can focus on our goals and life instead. Once you have experienced the capsule lifestyle, you may become quickly hooked, allowing capsule wardrobes to last for another fifty years.

For more information on capsule wardrobes and inspiration to get you started, here are a few resources to check out online:

10 Basics You Need in Your Closet for a Capsule Wardrobe | by Erin Elizabeth: https://www.youtube.com/watch?v=oDbBxJrl8R8

My Green Closet : Building a Sustainable Capsule Wardrobe and More https://www.youtube.com/channel/ UCxACkFpwxQpt6l9oNNR_Xcw

HOW-TO BUILD A CAPSULE WARDROBE: tips from a
stylist by Christie Ressel
https://www.youtube.com/watch?v=FFfOVosNIUw

Change Your Closet, Change Your Life | Gillian Dunn |
TEDxWhiteRock Inspiration on how organization and
minimizing can truly make an impact!
https://www.youtube.com/watch?v=WiVHSRY2I5Y

SPRING FASHION GUIDE | Trends & Capsule
Wardrobe Basics While this is an older guide to
spring fashion, Lilylikecom provides timeless tips on
transitioning and incorporating spring fashion looks
into your capsule wardrobe.
https://www.youtube.com/watch?v=4TU5YyKjm6U

Finding a Lookbook App

Any of these apps might be helpful for you to save images of
pieces and looks you would like to incorporate into your
capsule wardrobe. This can also help you keep your ideas
organized so you are more disciplined when shopping!
http://personallookbook.com/
https://www.flipsnack.com/digital-lookbook

Pinterest is also a great resource for "pinning" looks you love.
Here is an example of using Pinterest to create a fashion
lookbook for yourself:
https://www.pinterest.com/laniarakhma/fashion-l-o-o-k-b-o-o-k/

REVIEWS

Reviews and feedback help improve this book and the author. If you enjoy this book, we would greatly appreciate it if you could take a few moments to share your opinion and post a review on Amazon.